Stefan Muser

# Legendary Wristwatches

## from Audemars Piguet to Zenith

4880 Lower Valley Road • Atglen, PA 19310

ISBN: 978-0-7643-4957-7
Printed in China

Originally published as *Legendäre Armbanduhren Von Audemars Piguet bis Zenith* by Heel Verlag, Konigswinter, © 2010 Heel Verlag.

Translated from the German by Omicron Language Solutions.

Published by Schiffer Publishing, Ltd.
4880 Lower Valley Road
Atglen, PA 19310
Phone: (610) 593-1777; Fax: (610) 593-2002
E-mail: Info@schifferbooks.com

For our complete selection of fine books on this and related subjects, please visit our website at www.schifferbooks.com. You may also write for a free catalog.

This book may be purchased from the publisher. Please try your bookstore first.

We are always looking for people to write books on new and related subjects. If you have an idea for a book, please contact us at proposals@schifferbooks.com.

Schiffer Publishing's titles are available at special discounts for bulk purchases for sales promotions or premiums. Special editions, including personalized covers, corporate imprints, and excerpts can be created in large quantities for special needs. For more information, contact the publisher.

# Contents

# Dear Reader,

*Legendary Wristwatches* is a short introductory guide to the fascinating world of wristwatch collection. I have chosen some of the most beautiful pieces from our previous auctions, which will help retrace and evaluate one hundred years of wristwatch history.

Every decade of the 20th century had its own unique trends. Before the First World War there was still no prevailing consensus about owning a watch that was carried on the wrist rather than in the pocket. Equally diverse were the shapes and styles. In the twenties this experimental phase continued, but watches were now being made using a specialized technique with smaller "tailor-made" movements. To distance itself from the old pocket watch, watchmakers focused on rectangular, square, barrel-shaped, or asymmetrical watch cases in Art Déco style.

The 1930s marked a revolution in society and the sport watch experienced its first high point. Chronographs for determining personal best times were the norm and were worn demonstratively on the wrist. The fighter pilots of World War II, in contrast, wore a watch over their padded uniforms so that it would always be in their line of vision during flight maneuvers.

After the end of hostilities, people turned back to the finer things. Calendars, particularly those with moon phases, moved into the watch lover's field of view and also inspired watchmakers to create more delightful precision pieces. Simple yet elegant chronometers of different price ranges dominated the 1950s, but by no means at all did every young man own a wristwatch. Thus, production volume of the Swiss and German watch industries boomed. With the introduction of automatic winding in the early '60s, a new class of watches was established, and the diver's watch became the symbol of a mobile and sporting society. In the '70s the European watch industry eventually split into a progressive and a conservative camp, and both went through some hard times. The Japanese watch industry led a full-scale assault with their inexpensive quartz watches and forced the big German and Swiss watchmakers into a cutthroat price war.

The Swiss watch aristocracy, on the contrary, concentrated on fine, expensive luxury watches with top quality components and traditional technology. When the market was saturated with cheap used watches at the end of the '80s, people suddenly began to be interested in the wristwatch as a piece of jewelry and as an object of cultural value. Mechanical watches experienced an unexpected renaissance.

The watches pictured and described on the following pages are valued according to a simple and handy key. Since prices are subject to seasonal and cyclical fluctuations, we have devised a system with a maximum six stars

**Stefan Muser** *is owner and director of the internationally renowned auction house Auktionen Dr. Crott in Mannheim, Germany, which specializes in high-quality pocket watches, wall clocks, and wristwatches.*

★
Interesting and trouble-free entry level wristwatch
**< 1,000 € ($1,100)**

★★
Historically significant collector's item
**1,000-2,500 € ($1,100-$2,750)**

★★★
A classic of high historical value
**2,500-5,000 € ($2,750-$5,500)**

★★★★
Technically and stylistically superior
collector's item
**5,000-10,000 € ($5,500-$11,000)**

★★★★★
Rare, sought-after complicated watch or design
icon
**10,000-25,000 € ($11,000-$27,500)**

★★★★★★
Exceptionally rare delicacy for ambitious
collectors
**> 25,000 € ($27,500+)**

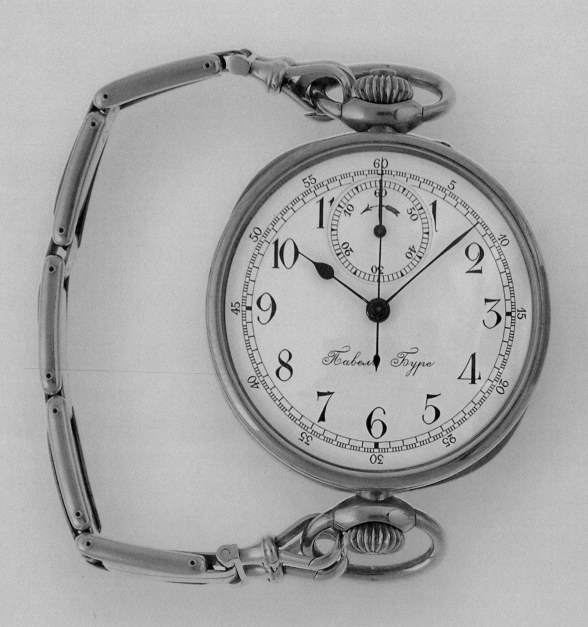

## Early Wristwatches

# From the Pocket to the Wrist

The first wristwatches came about in the period before World War I when coil loops were soldered to the housing of small, women's pocket watches. Soon after, the first watch case that was specifically constructed for mounting a leather strap arrived. But there was still a considerable difference between practical watches (for the military, for example) and fashionable time keepers.

## 1879

# GIRARD-PERREGAUX

**Wristwatch**

In 1879, German Chancellor Wilhelm I ordered a series of watches that could be "worn around the arm on a strap." This was done so that his officers didn't have to rummage around laboriously in their pockets to find the time. Museum piece (replica).

**Value:** ★ ★

**1915**

# OMEGA

**Deck Watch**

The Brandt brothers also manufactured wristwatches for military applications under their brand name Omega. To protect the sensitive glass, a grill could be attached. Museum piece.

**Value:** ★★★

**1912**

# PATEK PHILIPPE

## Rectangular Men's Watch

A very early men's watch by Geneva's manufacturer of elegant timepieces; with a contemporary rectangular watch case and an unusually sporty clock dial with gleaming numerals and hands. Exquisite hand-wound movement with mustache lever escapement. Gold case (18 karats); hand-wound.

**Value:** ★★★★★

## 1915

# OMEGA

**Men's Watch**

Early wristwatch by Omega, which was sold in Canada and Latin America under a newly created brand name: the Regina Precision Watch. Gold case (14 karats) with engraved cuvette, enameled dial; hand-wound.

**Value:** ★★

**1915**

# OMEGA

## Aviator's Watch with Chronograph

A remarkable witness to history: it was not only one of the first wristwatches but also one of the very first aviator chronographs. With a diameter of 46 mm, the clock was easy to read, and the solid pocket watch movement guaranteed precision. Nickel case with hinged case back; hand-wound.

**Value:** ★★★★★★

**1913**
PATEK PHILIPPE

**Curvex**

With curved case in 18-karat gold.

**Value:** ★★★★★

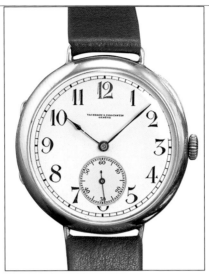

**1915**
VACHERON & CONSTANTIN

**Men's Watch**

Pocket watch case with soldered watch band brackets.

**Value:** ★★★

**1915**
LONGINES

**Men's Watch**

Pillow-shaped case with enameled dial.

**Value:** ★★

**1918**
IWC

**Men's Watch**

Silver case with soldered watch band brackets.

**Value:** ★★★

**1920**
ULYSSE NARDIN

**Chronograph**

Very early crown-pusher chronograph; pocket watch case.

**Value:** ★★★

**1924**
CARTIER

**Santos**

The first wristwatch constructed for a specific purpose (1904).

**Value:** ★★★★

14

## Art Deco

# An Idea Catches On

In the "Roaring Twenties" the wristwatch began to gain in popularity over the still prevalent pocket watch. To distinguish itself from those old-fashioned time keepers, manufacturers devised rectangular, oval, and trapezoidal shaped watch cases. Movements were made with parts that were completely new and smaller, and they were often rectangular or barrel-shaped themselves.

**1926**

# ROLEX

## Prince Classic Chronometer Extra Prima Ref. 1862

With two neatly separated dials (time and auxiliary seconds), the Rolex "Prince" was recommended for doctors since the uniquely arranged seconds hand made pulse measuring easier. Rolex always emphasized the importance of clock accuracy, which is why many watches were delivered as certified chronometers. Gold case (9 karats); hand-wound.

**Value:** ★★★

**1927**

# PATEK PHILIPPE

## Square Men's Watch

In the 1920s, the precious metal platinum experienced a real boom. It was especially prized for its cool, blueish shimmering brilliance, which was quite fitting in this era of technological advancements. Patek Philippe combined the rigid, square shape of the case with curved "Breguet" numerals. Platinum case with hinged case back; hand-wound.

**Value:** ★★★★

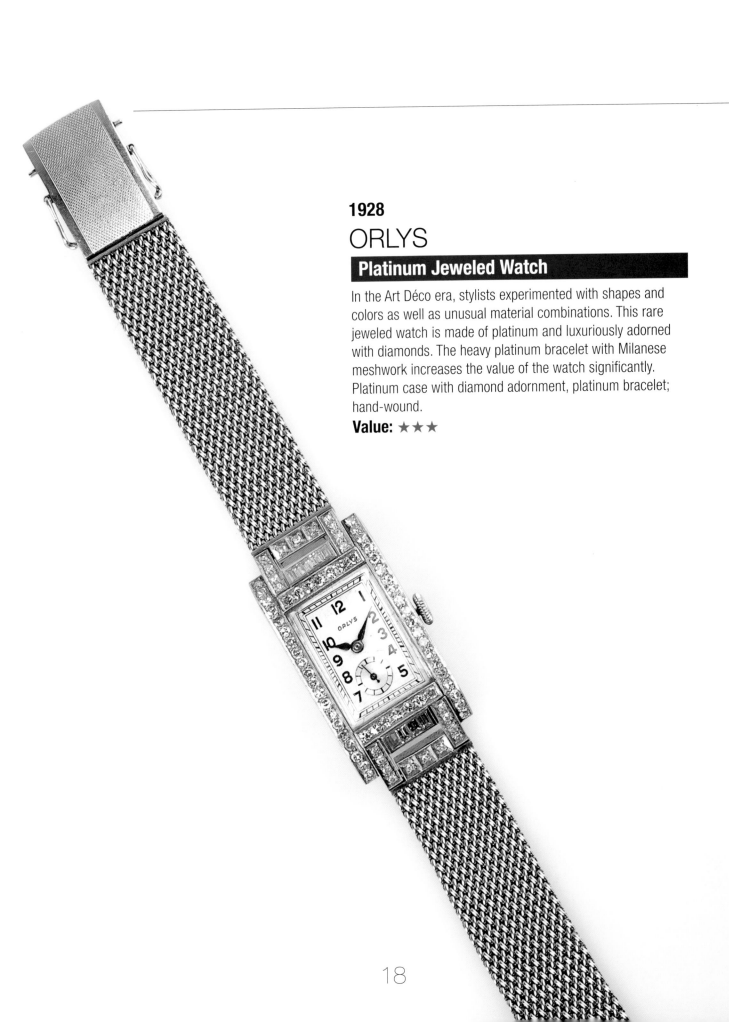

**1928**

# ORLYS

## Platinum Jeweled Watch

In the Art Déco era, stylists experimented with shapes and colors as well as unusual material combinations. This rare jeweled watch is made of platinum and luxuriously adorned with diamonds. The heavy platinum bracelet with Milanese meshwork increases the value of the watch significantly. Platinum case with diamond adornment, platinum bracelet; hand-wound.

**Value:** ★★★

**1928**

# PATEK PHILIPPE

**Gondolo**

Fine rectangular men's watch with hand-wound rectangular movement. The white gold case (18 karats) is uncommon; the curved Breguet numerals are typical.

**Value:** ★★★★★

# Patek Philippe

The story of **Patek Philippe** begins with a Polish count named Antoine Norbert de Patek. He sought political asylum in Geneva and once there discovered a love for fine watchmaking. Together with his fellow countryman François Czapek, a talented watchmaker, the former officer founded the company Patek, Czapek and Co. in 1839. With six employees, they produced about 200 watches per year. The aesthetic and technical qualities of these watches assured them an enviable position from the beginning on.

Patek was holding an exhibit of his products in Paris when he first heard of watchmaker Jean-Adrien Philippe. Philippe had designed a flat pocket watch that could be wound and set by the crown — an innovative improvement of great significance. Because Patek's contract with Czapek had expired, he offered to collaborate with Philippe, and on May 15, 1845, the company was renamed Patek & Co. But since Patek owed the success of his watches to his brilliant watchmaker Philippe, he incorporated his name into the company. And so beginning in 1851, it was known as Patek Philippe & Co. They produced their first wristwatch in 1868 for the Hungarian Countess Kocevicz — it was probably even the first wristwatch in the history of Swiss watchmaking.

The death of Antoine de Patek in March 1877 could not slow down the innovative drive of Jean-Adrien Philippe. In 1889, he constructed a movement with a "perpetual" calendar date, whose display required no adjustment whatsoever — not for the varying length of months or for leap years. When Jean-Adrien Philippe passed away in 1894, his stepson Joseph-Antoine Benassy-Philippe took control of the business, and seven years later, he converted it into an incorporated company.

The Philippe family had to sell the Patek Philippe factory as a consequence of the global economic crisis in 1929, but the company remains a family business to this day. The Stern brothers, still the exclusive dial distributors of Patek Philippe, bought the factory in 1932 and expanded it. Under their aegis, the finest wristwatches the world had ever seen were being produced in Geneva: watches with numerous different complications (from perpetual calendars over chronographs to tourbillons) as well as quite simple, timeless models of extremely exquisite workmanship. In the world of wristwatch collecting, Patek Philippe watches from this era are very highly valued. In fact, the most expensive wristwatch ever auctioned is, of course, a Patek Philippe.

**1928**

# PATEK PHILIPPE

**Rectangular Men's Watch**

Fine gold rectangular watch from Patek Philippe's Geneva factory. Gold case with hinged case back; hand-wound.

**Value:** ★★★★★

**1929**

# PATEK PHILIPPE

## Rectangular Men's Watch

Two-toned watch cases were very fashionable in the 1920s. Patek Philippe also produced such a model, featuring a rigid white gold case softened by long rose gold sides. White/rose gold case (18 karats); hand-wound.

**Value:** ★★★★

**1930**

# MOVADO

## "Polyplan" Men's Watch

Thanks to a double-angled movement plate, the Polyplan's case could be bent enough so that it fit snugly against the wrist. The assembly of the movement was very complex and expensive and production of the watch ceased before the Second World War. Gold case (18 karats); hand-wound.

**Value:** ★★★★

**1929**

# ROLEX

**Prince Brancard Chronometer Extra Prima Ref. 971**

The so-called "Brancard" shape can be identified by the slim waistline of its elongated case. Especially coveted are the "striped" versions with two gold colors. Gold/white gold case; hand-wound.

**Value:** ★★★★★

**1929**

# PATEK PHILIPPE

## "Officier"

Elegant pillow-shaped gold watch with an exquisite hand-wound movement. Gold case (18 karats), enameled dial; hand-wound.

**Value:** ★★★★

**1920**

PATEK PHILIPPE

**Men's Watch**

One of the brand's early wristwatches with an unusual case.

**Value:** ★★★★★

**1920**

IWC

**Men's Watch**

Tasteful Art Déco case with painted black decor.

**Value:** ★★★★

**1920**

CARTIER

**Tortue**

Early wristwatch in a typical "tortoise" case.

**Value:** ★★★★

**1924**

ROLEX

**Oyster "For All Climates"**

Very early waterproof wristwatch of sturdy construction.

**Value:** ★★★

**1925**

LONGINES

**Men's Watch**

Large rectangular wristwatch with bowed case.

**Value:** ★★

**1925**

VULCAIN

**Men's Watch**

Rare oval wristwatch with arched case.

**Value:** ★

**1925**

CARTIER

**Tank Cintré**

Elongated, slightly bowed "Tank"-shaped case in gold.

**Value:** ★★★★★

**1928**

A. LANGE & SÖHNE

**Men's Watch**

One of the few wristwatches by Lange from the 1920s.

**Value:** ★★★★★

28

# Chronographs of the 1920s to 1950s
# A Sporting Challenge

One of the first additional wristwatch functions to attain great popularity was the chronograph. The installation of a stopwatch into an existing movement is quite complicated and costly. Luckily these watches turned out to be a bit bigger so watchmakers could revert back to pocket watch chronograph movements.

**1920**

# BUHRÉ

## Chronograph

A remarkable early chronograph wristwatch, based on the pocket watch. The narrow bracelet was mounted on the protective brackets of the crowns, rendering the second one inoperable. Gold case (14 karats) with gold bracelet; hand-wound.

**Value:** ★★★

**1920**

# H. MOSER & CIE.

## Chronograph

Like the Buhré watch pictured on the previous page, this wristwatch was intended for the Russian market, where Moser was strongly represented. These chronographs already included a minute counter. Gold case (14 karats) with gold bracelet; hand-wound.

**Value:** ★★★

**1922**

# LONGINES

## Chronographe à Compteurs

An early chronograph by Longines; rare and finely crafted. Typically, the first chronographs had only a single pusher, and oftentimes (as shown here) it was housed in the hollow crown. Only successive timings were possible with it (start, stop, and reset), which meant that a measurement could not be interrupted and recorded again. Gold case (18 karats) with hinged case back, enameled dial; hand-wound.

**Value:** ★★★★

**1928**

# ROLEX

## Chronograph Ref. 2508

Long before the Daytona left its mark on chronograph history, Rolex offered its customers a sporty yet elegant stopwatch. This model is noteworthy not only due to its gold case, but also because of its black dial. Gold case (18 karats); hand-wound.

**Value:** ★★★★★

**1932**

# ROLEX

**Chronograph "Anti-magnetic" Ref. 2508**

This sporty Rolex chronograph from the 1930s is known among horologists as Reference 2508. The first models still had oval pushers, but the new case received two rectangular pushers: one for starting and stopping the timing device and one for resetting the hand, which made determining intervals possible. Special edition for Juwelier Bucherer, Lucerne. Stainless steel case; hand-wound.

**Value:** ★★★★★

**1935**

# UNIVERSAL GENÈVE

**Chronograph**

A fine gold chronograph by Universal Geneva with rectangular case — typical for the 1930s. The elegant silver-plated dial is printed with a sophisticated scaling and affixed with thin hands. Rose gold case (18 karats); hand-wound.

**Value:** ★★★

**1939**

# PATEK PHILIPPE
## Chronograph Rattrapante Ref. 1436

Extremely rare, exquisite chronograph with a split seconds function, or double chronograph. One of the two superimposed second hands can be stopped temporarily, to mark intervals for example. When started again, it closes in on its "colleague," which in the meantime carries on running.

**Value:** ★★★★★★

This tremendously complex
hand-wound movement has a
diameter of 30 mm and comes
with two ratchet wheels for driving
the chronograph functions. The
column wheel, which can hold and
release the split second hand, can
be seen in the movement's center.

# Longines

The name **Longines** derives from an entry in the municipal land registry office of St. Imier: the company's headquarters are in the meadows that line the River Suze. In French, the area is known as "Les Longines," or "the long meadows."

Agassiz & Co. was founded in 1832 by Auguste Agassiz and his partners Henri Raiguel and Florian Morel. It began as a piecework operation, in which complete watches were built from purchased components. The assembly mainly took place in homes, while Agassiz & Co. specialized in sales.

In 1852 Auguste Agassiz relinquished control to his nephew Ernest Francillon, who recognized that inexpensive, sturdy watches were in high demand. Francillon decided he had to streamline his watch production. In 1866, he acquired the above-mentioned plot of land and quickly made the meadow's name the brand name of his watches. Because the river delivered energy free of charge, he was able to develop mechanically supported production methods in the newly built factory. Francillon delivered low-cost movements and watches with consistently high quality, and soon he would export his watches to the entire world.

Longines became synonymous with professional aviator watches thanks to US Postal Service pilot Charles A. Lindbergh, who had become the first to fly solo across the Atlantic on May 21, 1927. Lindbergh sketched his idea for a practical aviator's watch with navigational aids and sent it to John P. V. Heinmüller, then director of Longines Wittnauer Watch Co. in the United States. Heinmüller was also president of the World Air Sports Federation and was an avid aviator himself. He recognized the value of Lindbergh's idea and contacted Longines in Switzerland, where the Lindbergh - Hour Angle Watch was prepared to go into production. From

then on, the watches from St. Imier became the preferred equipment of pilots and explorers. Pioneers of flight, like Amelia Earhart, Howard Hughes, and Richard Evelyn Byrd relied on their Longines watches.

By 1912 Longines had introduced an electromechanical system for activating and stopping chronographs, which qualified the company to be an official timekeeper of big sporting events and the Olympic Games.

Like many other brands, Longines lost its independence during the quartz revolution. Initially it aligned itself with other smaller brands to form ASUAG, which later became the Swatch Group after further mergers and name changes.

**1940**

# LONGINES

**Chronograph "Anti-Magnétique"**

This rather small men's watch (34 mm in diameter) with chronograph has a classically shaped dial with a tachometer scale. Stainless steel, push back; hand-wound.

**Value:** ★★★★

**1940**

# VACHERON & CONSTANTIN

## "Doctor's Chronograph"

This rare chronograph in a pillow-shaped case comes with a pulsometer scale, which makes a doctor's work easier — instead of counting every heartbeat for a minute long, counting 30 beats suffices to calculate the projected pulse indicated by the pulsometer scale. Silver case; hand-wound.

**Value:** ★★★★★★

**1940**

# VACHERON & CONSTANTIN

## Chronograph

Rare, fine chronograph in stainless steel case with two rectangular pushers (start/stop and reset). The dial is bordered with a tachometer scale. Stainless steel case; hand-wound.

**Value:** ★★★★★

**1940**

# LONGINES

**Chronograph**

A fine chronograph with 12-hour counter and central 60-minute counter with red hand. The famous Longines 13ZN caliber served as its basis. Stainless steel case with screw back; hand-wound.

**Value:** ★★★★★

**1940**

# JAEGER

## Chronograph

A rare Jaeger chronograph (without LeCoultre) with blue tachometer scale and two concentric telemeter scales (red and blue) in heavy gold case (18 karats). Hand-wound.

**Value:** ★★★

**1944**

# OMEGA

## Chronograph

This fine chronograph comes with an additional telemeter scale, which calculates the dispersion rate of sound waves in kilometers. This feature was originally invented for the military, which could determine the distance of a shot from the time difference between the muzzle flash and the gun blast. In the civilian world, the distance of an approaching storm could be deduced. Stainless steel case; hand-wound.

**Value:** ★★★★

**1945**

# LONGINES

## Chronograph

Very fine gold chronograph in a massive rose gold case with an uncommon two-color dial. Typical rectangular pushers and contoured lugs in the style of the period. Rose gold case (18 karats); hand-wound.

**Value:** ★★★★

## 1945

# OMEGA

**Chronograph**

This typical prewar chronograph has a hand-wound movement with a small permanent seconds counter and a symmetrically arranged 30-minute counter. The stopwatch hand emanates from the center and the red tachometer scale is especially striking. Stainless steel case with screw back; hand-wound.

**Value:** ★★★★★

**1945**

# MIDO

## Multicenter Chrono

A very rare "doctor's chronograph" design with a pulsometer scale (calculated by 30 heartbeats). The second hand and minute counter of the famous Multicenter Chrono really jump out from the dial's center. Refined stainless steel case with red gold covering. A soft iron cage shields the movement from magnetic fields. Screw back; hand-wound.

**Value:** ★★★

A finely finished chronograph caliber;
classic layout with ratchet wheel steering
and screw balance.

### 1950
# LÉMANIA
**Chronograph**

This simple chronograph from the Italian Hydrographic Institute's inventory has only a single pusher, which allows for simple time measurements. Stainless steel case with screw back; hand-wound.

**Value:** ★★

**1925**
UNIVERSAL GENÈVE
**Chronograph**

Gold case, enameled dial, crown pusher.
**Value:** ★★★

**1925**
EBERHARD & CO.
**Chronograph**

Single pusher, enameled dial.
**Value:** ★★★

**1930**
LONGINES
**"Doctor's Chronograph"**

Enameled dial, equipped with pulsometer scale.
**Value:** ★★★★

**1935**
BREITLING
**Chronograph**

Improved with two pushers: start/stop and reset.
**Value:** ★★

**1937**
FORTIS
**Chronograph "Wandfluh"**

Enclosed case housing with a movement container that is pressed in from behind.
**Value:** ★

**1945**
HEUER
**Chronograph**

Sporty stopwatch in polished stainless steel case.
**Value:** ★★

**1947**

# PATEK PHILIPPE

## Chronograph Ref. 130

Highly coveted by collectors, Reference 130 was first produced in 1934 and it came in a stainless steel case or in an optional gold, or rose gold, case. Only three white gold copies are known to exist. The skillfully finished movement is based on an ébauche from Lémania. Stainless steel case; hand-wound.

**Value:** ★★★★★★

Patek Philippe caliber 12'''-120.
Horologists all agree — to this day,
there is no chronograph movement more
beautiful or more valuable.

53

**1959**

# PATEK PHILIPPE

## Chronograph Ref. 1463

This model, also known as Reference 1463, was first produced in 1940. Technically and stylistically, it embodies the quintessence of excellent Swiss chronograph construction from the era before World War II. Extremely rare in stainless steel case, and even more so with Breguet numerals! With soft iron cage to shield against magnetic fields. Screw back; hand-wound.

**Value:** ★★★★★

This gorgeous hand-wound movement (Caliber 12'''-120) is hidden under an additional magnetic field protection cap and only reveals itself to the watchmaker.

Watches of the 1930s

# Elegant and Sporty

Two social currents influenced the development of the wristwatch in the '30s: first of all, a watch worn on the wrist proceeded to become more and more of a real status symbol, and so new watch designs became progressively more elaborate. Secondly, the wristwatch had to confront watch wearers' increasing dedication to leisure sports and become considerably more durable.

**1935**

# PATEK PHILIPPE

## Rectangular Men's Watch Ref. 417

Fine men's watch with the legendary caliber 9-90 rectangular movement.
Intricate dial with affixed numerals and baton indexes with distinctive auxiliary
"small seconds" sub-dial. Stainless steel; hand-wound.

**Value:** ★★★★★

**1938**

# PATEK PHILIPPE

## Men's Watch Ref. 564

Exceptional rectangular watch with tiered gold case and hidden lugs. Only five copies were manufactured for the jeweler Mappin & Webb in Rio de Janeiro. Rose gold case (18 karats); hand-wound.

**Value:** ★★★★★

**1939**

# PATEK PHILIPPE

## Rectangular Men's Watch Ref. 425

Exquisite men's watch in heavy 18-karat red gold case with red gold hands and baton indexes. Handwound.

**Value:** ★★★★

**1939**

# PATEK PHILIPPE

**Rectangular Men's Watch**

Prominent, ball-shaped decorations an all four lugs give this watch a distinctive look. Rose gold case (18 karat); hand-wound.

**Value:** ★★★★

# Portrait

# Rolex

**Rolex** is unrivaled in the watch industry. The brand long ago said goodbye to all conventional standards, and their products gained a mythical reputation. Rolex stands for glamour and wealth, but also for precision and innovation. And this is all attributed to one man: Hans Wilsdorf from Kulmbach, Bavaria.

This self-made man — he became an orphan at 12 years of age and had to make his way in the world alone — first learned about chronometers in a large watch trading company in La Chaux-de-Fonds. He later took his experience to London and built his own watch trading office with his wealthy partner Davis. He then turned his attention to wristwatches, as these models had been in demand with men ever since the Boer Wars — they had proved to be quite useful in Africa and also carried a "manly" image.

Wilsdorf relied on small movements, which he obtained from Aegler, a Swiss manufacturer in Biel. In clock accuracy and durability, they were superior even to some pocket watches. In 1910, an Aegler caliber earned a chronometer certificate, the first for a wristwatch movement.

The Rolex brand name, together with its five-pronged crown, had already successfully established itself on the market when, in 1919, Wilsdorf had to relocate his business to Switzerland due to tariff problems in England. In Geneva, he founded the company Montres Rolex S.A., and the movement factory in Biel was bought out and became "Rolex Watch Co., Aegler S.A."

Rolex owes its success to two big inventions that gave the wristwatch leverage and made it what it is today: the waterproof watch and the self-winding watch. Wilsdorf had already developed the absolutely waterproof "Oyster" case in the 1920s. But because the movement's constant winding presented a danger to the impermeability of the crown, Wilsdorf designed the first automatic rotor ready

for production, and in 1931 the Oyster Perpetual was born. From then on, the Oyster set the standard and most automatic watches today work according to its principles.

The Oyster is still the company's most well-known model, more popular than all the diver's watches and chronographs. It's even more popular than the early Prince models and the Cellini, which were not only dependable watches but elegant too.

Thanks to consistent advancements, the Oyster remained the benchmark for rugged sport's watches. Modified as the Submariner, it became the first professional diver's watch. It was even fitted with a helium valve for professional dives requiring special gas mixtures.

Rolex watches also proved their impressive dependability on several expeditions. They have been on Mount Everest, at both the North and South Poles, at the deepest point on earth (the 36,000-foot-deep Mariana Trench) and even in space.

Approaching old age and without any children, Wilsdorf transferred his stock to the Wilsdorf Foundation, which he established. After his death in 1960, André J. Heiniger took control of the business and his son Patrick Heiniger succeeded him as president in 1992. In 2008, the family dynasty came to an end. Since then, the five- person foundation board has governed the company.

**1939**

# PATEK PHILIPPE
## Men's Watch Ref. 1426

Exquisite men's watch with hidden watch band lugs, called "hooded lugs" in collector's jargon. The two-toned gold dial with affixed hour markers is eye-catching. Rose gold case (18 karats); hand-wound.

**Value:** ★★★★★

**1938**

# PATEK PHILIPPE

**Men's Watch Ref. 1491**

Beautiful example of a round men's watch with unique
scroll lugs. Rose gold case (18 karats); hand-wound.
**Value:** ★★★★★

**1934**

# OMEGA

## Marine

The first waterproof rectangular watch came with an elaborate sliding outer case and locking clip. This model was first sold in 1932. Uncommon in gold case (14 karats), specially produced for the Canadian jeweler Birks; hand-wound.

**Value:** ★★★

**1938**

# ROLEX

## Oyster Perpetual Chronometer Ref. 3064

The Rolex Oyster was way ahead of its time with an automatic movement in a hermetically sealed case and screwed-down crown. The so-called "Bubble Back" was the prototype for the modern wristwatch. Stainless steel/gold case and bracelet; self-winding.

**Value:** ★★★★★

**1938**

# IWC

**Men's Watch**

This no-frills men's watch was manufactured by
IWC for the jeweler Türler, as noted on the dial.
Stainless steel; hand-wound.

**Value:** ★★★

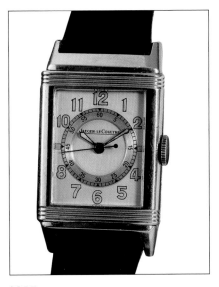

**1932**
AUDEMARS PIGUET
**Digital Display**
Rare white gold watch with digital
(discs) time display.
**Value:** ★★★★★

**1935**
JAEGER-LECOULTRE
**Reverso**
The rectangular case could be turned
over to protect the glass.
**Value:** ★★★

**1935**
TISSOT
**Men's Watch**
Typical rectangular men's watch with
two-colored dial.
**Value:** ★

**1938**
CARTIER
**Tank "Louis Cartier"**
Golden "Tank" watch with a beveled
emerald in the crown.
**Value:** ★★★

**1936**
JUNGHANS
**Men's Watch**
Robust rectangular watch with simple
but solid construction.
**Value:** ★

**1935**
DOXA
**Men's Watch**
Elegant, simple men's watch with mov-
able lugs. Stainless steel; hand-wound.
**Value:** ★

Aviator Watches

# Precision Above the Clouds

With arms production booming in Europe during the 1930s, the watch industry made enormous technological improvements. Aviator watches and chronographs of this era represented the qualitative peak for the industry. But because the military defined many of the dial and watch case designs, there was a lack of aesthetic diversity.

**1934**

# LONGINES WITTNAUER

## "Lindbergh" Hour Angle Watch

The Hour Angle navigation watch was designed according to Charles A. Lindbergh's invention; equipped with rotating bezel and adjustable center dial. Stainless steel case with hinged case back; hand-wound.

**Value:** ★★★★★

**1938**

# LONGINES

## "Weems Navigation" Watch

This watch, named after an invention by Captain Van Horn Weems, has a rotating bezel with 60-minute graduation that can also be secured in place.

**Value:** ★★★

LONGINES
WEEMS

U.S. PAT. 2008734

SWISS MADE

**1940**

# MOVADO

**Aviator Navigation Watch**

The bezel, which is held in place by a second crown to avoid unintentional displacement, lets you set bezel markings or synchronize the second hand with a time signal. Stainless steel case; hand-wound.

**Value:** ★★

**1936**

# OMEGA

## Aviator Watch

This early aviator watch has a rotating glass ring
with index markers for setting bezel markings.
Stainless steel case: hand-wound.

**Value:** ★★★★

**1936**

# IWC

## "Mark IX" Aviator Watch

Early on, the International Watch Co. in Schaffhausen specialized in the production of wristwatches for use in an airplane cockpit. The Mark IX was the beginning of continuous development. Stainless steel case, rotating bezel, magnetic field protection; hand-wound.

**Value:** ★★★★★

Four large aviator watches used by the German Air Force during World War II. Above and far right are two examples from A. Lange & Söhne of Glashütte, Germany. In the center (top) is a watch from Walter Storz (Stowa), and below it one from Lacher & Co. (Laco), both of Pforzheim.

**1940**

# IWC

## Large Aviator Watch

After a military request for proposal, IWC produced several thousand aviator watches for the German Air Force. Because of their outstanding quality, they are highly valued today. The case of these "large" aviator watches has a diameter of 55 mm. Stainless steel; hand-wound.

**Value:** ★★★★★

**1945**

# PANERAI

## Radiomir

Not an aviator watch, but an Italian diving watch produced for the German Navy. "Radiomir" denotes the weak radioactive glowing substance that illuminates the numerals and hands. Waterproof wristwatches for professional or military dive missions were sparse in the 1940s. Stainless steel with screw back; hand-wound.

**Value:** ★★★★★★

## Portrait

# IWC

When 27-year-old Florentine Ariosto Jones, an American engineer from Boston, founded the International Watch Company in Switzerland, he did so mainly because Switzerland was a relatively low-wage country in the mid-19th century in comparison to the USA. The heavily mechanized watch industry in America had also reached the limits of its capacity. During his search for new production sites, Switzerland, with its watchmaking know-how and low-wage levels, appealed to Jones. In the Swiss town of Jura, where most of the watch companies were, nobody wanted to hear about machine automated production. But Jones received an offer from the city of Schaffhausen on the Rhine River and he also found a potential business partner named Johann Heinrich Moser. Moser had invented a new system that involved using the power of the fast flowing waters of the Rhine for industrial purposes. Water wheels would activate cog

wheels and transmission wires, which in turn could dispense power to machines. Thanks to Moser's invention, the new watch company had all of the necessary energy at their disposal, and practically free of charge. Jones found this agreeable and Moser became cofounder of the International Watch Company in 1869. Just one year later, IWC would produce 10,000 watches — all for the American market.

But these exports came to a halt in 1874 when the USA began demanding a 25 percent protective tariff on foreign wares. The owners then converted IWC into an incorporated company. Jones returned to Boston in 1876 and IWC was bought out by Schaffhauser Handelsbank.

Four years later the entrepreneurial Rauschenbach family took over the reins.

They opened up new markets for IWC — first and foremost in neighboring Germany, where their watches were in high demand thanks to the outstanding quality. IWC's product line geared itself toward market trends and so wristwatches also began to appear in the window displays of concessionaires.

They later recognized the importance of aviation and began offering watches that could be worn over a pilot's thick flight jacket. The tradition of IWC aviator watches continues to this day.

It would be unfair, though, to label IWC as solely a producer of aviator watches. In fact, all of IWC's movements are considered exceptionally well constructed and technically sophisticated — not only in collector circles. IWC also took its own approach to watch case construction and in 1954 the "Ingenieur" was launched. In its watch cases, a soft iron cage protected the movement from magnetic interference, which normally affected a watch's accuracy. The simple models of the 1930s, manufactured in Schaffhausen, are also renowned. They embody a modest understatement backed up by high quality.

In the 1970s IWC, together with Jaeger-LeCoultre, was sold to the VDO/Mannesmann Group, which became the Richemont Group in 2000.

**1942**

# LACO

## Aviator Watch

German Air Force aviator watch produced by Lacher &
Co. in Pforzheim after a military request for proposal.
Stainless steel case; hand-wound Durowe caliber.

**Value:** ★★★

## Calendar and Moon Phases

# Small Complications

In the postwar period, wristwatches with date functions and elaborate calendar displays became popular. Even the depiction of the moon, which already existed in pocket watches, was discovered again. Particularly complex movements combined these displays with chronograph functions.

**1920, 1924**

# HEINRICH MOSER & CIE.

## Calendar Watches

Two examples of very early wristwatches with calendar displays from Heinrich "Henry" Moser of Neuhausen, Switzerland. The central, movable lugs are soldered to small pocket watch cases (33 – 34 mm diameter). The calendar displays comprise a separate hand indicating the date and a weekday aperture. Nickel and red gold case (14 karats) with hinged case back; hand-wound.

**Value:** ★★★

# Jaeger-LeCoultre

Actually, the company's name could have been written the other way around since the story of **Jaeger-LeCoultre** begins with Antoine LeCoultre. Antoine was born in 1803 in Le Sentier, in the Swiss Vallée de Joux (Valley of Watches), where his family roots date back to the 16th century. At 30 years of age, he founded a manufacturing plant for movement parts in his home town, where the Jaeger-LeCoultre factory is still located today.

Antoine LeCoultre was quite an innovative man. In 1844, he contrived the "Millionometer" — an instrument that was capable of measuring to the one one-thousandth of a millimeter (one one-millionth of a meter). Today's double function of the crown — winding the movement and setting the hands — can be traced back to LeCoultre's invention, which made winding by key obsolete. LeCoultre owned several patents, and after his death, the LeCoultre & Compagnie factory specialized in the production of fine, complicated movements.

The second part of the story began in 1925 with the entrance of Alsatian engineer Edmond Jaeger (born 1850), who had previously supplied Cartier, the French Navy, and others with movements. Manufacturing opened up the possibility for him to develop and produce not just movements, but also complete wristwatches under the double logo. In 1926, the technically interesting "Duoplan" emerged, followed in 1928 by the table clock "Atmos," which obtained its operating power from the temperature differences of its surroundings.

In 1931 the "Reverso" appeared in the product line for the first time. With a simple flick of the finger, its rectangular case could be turned over to protect the watch's delicate glass. According to legend, the idea for the extraordinary case design, which is still produced in countless variations today, was conceived by British polo players who were on duty in India.

Jaeger-LeCoultre soon enjoyed an outstanding reputation for capable craftsmanship. Among their specialties, then as well as today, are calendar watches with moon phases, perpetual calendars, and alarm watches. In 1948, after the death of Jaques-David LeCoultre, ownership of the business changed over to SAPIC Holding. The company went through several hands before it landed under the roof of VDO/Mannesmann at the end of the 1970s. By the late '90s the group sold their entire watch segment ("Les Manufactures Horlogères" with Jaeger-LeCoultre, IWC, and A. Lange & Söhne) to the Richemont Group.

**1945**

# ETERNA

## Chronograph with Full Calendar

Elegant chronograph with calendar displays.
Gold case (18 karats); hand-wound.

**Value:** ★★★★

**1949**

# ROLEX

## Chronograph Dato Compax Ref. 4768

Rare, sporty men's watch with chronograph and full calendar (date, day, and month displays). Stainless steel case; hand-wound.

**Value:** ★★★★★★

**1949**

# OMEGA

## "Cosmic" Calendar Watch

Rare, fine full calendar watch with moon phase display in exceptionally good, original condition. Rose gold case; hand-wound.

**Value:** ★★★

**1947**

# JAEGER-LECOULTRE

## Calendar Watch

Full calendar watch with apertures for day and month in heavy gold case with unusually shaped lugs. Gold case; hand-wound.

**Value:** ★★★

**1953**

# ROLEX

**Oyster Perpetual Chronometer Ref. 6062**

One of the most important and rarest chronometer models from Rolex with full calendar (date, day, month) and moon phase display. Gold case and bracelet (18-karats); self-winding.

**Value:** ★★★★★

**1949**

# UNIVERSAL GENÈVE

**Full Calendar with Moon Phases**

The unusual dial layout demands a second look: the date is by the "3," day display by the "12," month and moon phases near the "6." Stainless steel case; hand-wound.

**Value:** ★★

**1949**

# AGASSIZ WATCH CO.

**World Clock**

Very fine and rare men's watch with world time display, which was actually Patek Philippe's domain at that time. Elegant gold case with flashy lugs. Gold case (14 karats); hand-wound.

**Value:** ★★★★

# On the Way to Automatic
# Elegance in the Wirtschaftswunder

Kidney-shaped coffee tables and champagne glasses: the Wirtschaftswunder (economic miracle) propelled the luxury goods industry. After Rolex's patent for the self-winding rotor expired, all of the Swiss watch manufacturers launched into automatic watches and cultivated the technology to the highest level.

**1952**

# PATEK PHILIPPE

**Men's Watch Ref. 2461**

Extremely rare, exquisite men's watch in a solid
rectangular case made of platinum. The valuable
precious metal experienced a mini-renaissance in the
'50s. Platinum case; hand-wound.

**Value:** ★★★★★

**1954**

# PATEK PHILIPPE

## Ref. 1593 "Hourglass"

Because of the case's prominent curving, this elegant men's watch is referred to as the "Hourglass" among collectors. This is a valuable platinum model. Hand-wound.

**Value:** ★★★★★★

**1955**

# VACHERON & CONSTANTIN

**"Cioccolatone"**

Due to the case's characteristic shape, this elegant men's watch is also called the "little piece of chocolate." Gold case (18 karats) and Milanese mesh bracelet; hand-wound.

**Value:** ★★★★★

**1955**

# BREGUET

**Men's Watch**

Uncommon white gold design of this rare men's watch.
Breguet, the old, steeped-in-tradition brand name, was
reduced to the size of a specialty workshop in the 1920s.
White gold case (18 karats); hand-wound.
**Value:** ★★★★★★

# Vacheron Constantin

**Vacheron & Constantin** can boast about being the oldest continuously active watch manufacturer. Their story began in 1755 when Jean-Marc Constantin went into business for himself in Geneva. He found customers for his sophisticated watches in France. That such an important market had emerged there was due to the spoiled and pompous noble class — Louis XIV himself was an enthusiastic watch collector. But in 1789 this market collapsed — the French Revolution kept the world on the edge of its seat and the aristocrats were busy saving their lustrous lives. Watches were no longer important, at least for the time being.

Constantin had fought hard to keep his company afloat. After he died in 1805 and the business was taken over by his heirs, they had no other choice but to search for a new, solvent partner. In 1819 they found one in François Vacheron, the son of a wealthy fabrics and grains merchant. He took control of the commercial side of the business, which was renamed Vacheron & Constantin.

The company's biggest success back then was owed to its watchmaker George-Auguste Leschot, who built a machine in 1839 that could produce precision movements rather economically. This initial step in mechanized fabrication was a small revolution for the watch industry.

Wristwatches were absorbed into the company's program around 1910, with movements that usually came from LeCoultre. Early on, people had a weakness for luxurious watches. In addition to el-

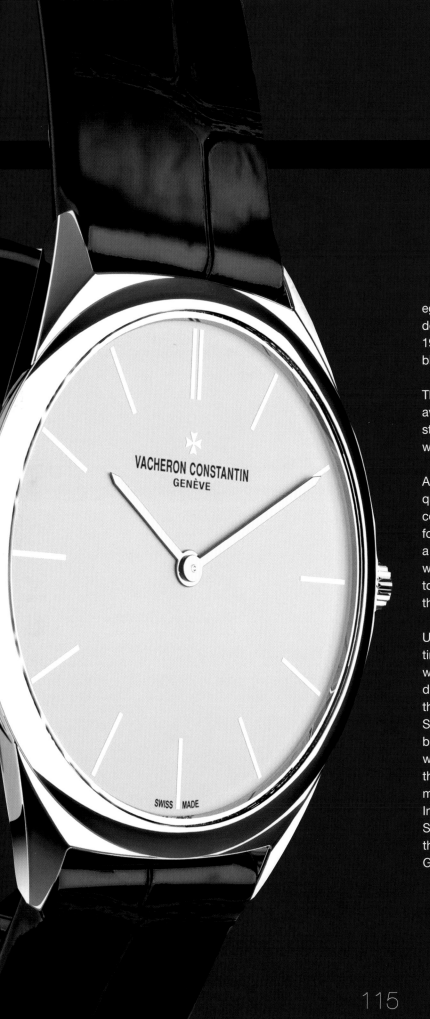

egant watches for everyday use, the watchmaker designed models with an avant-garde look, in the 1920s, a designer watch was built with a small blind to cover the clock face.

There was also a big selection of chronographs available beginning in the 1930s. Vacheron & Constantin offered a wristwatch with minute repeating, which was only available in a limited series.

Although Vacheron & Constantin produced high-quality watches, the owners failed to keep the company financially independent. After they again found themselves in a financial crisis — this time as a consequence of the Second World War — there was nothing left for Charles Constantin to do but to sell his majority stake. Georges Ketterer became the new owner.

Under Ketterer's leadership, Vacheron & Constantin cemented its position as a producer of luxury watches. But even then, most of their movements didn't originate from their own factory. Instead they were purchased from Jaeger-LeCoultre in Le Sentier. In 1955, as part of the celebrations for the brand's 200th anniversary, an ultra-thin wristwatch was introduced. Its movement was only 1.64 mm thick — still among the thinnest movements ever made — and is still produced in limited quantities. In the 1980s, an investment firm, led by the Arabian Sheik Yamani, purchased the company. In 1996, the Vendôme Group (today the Richemont Luxury Group) made him an offer he couldn't refuse...

**1950**

# BREITLING

## Duograph

Rare and very elegant split-second chronograph in heavy red gold case. Hand-wound.

**Value:** ★★★★

**1957**

# PATEK PHILIPPE

## Men's Watch Ref. 2551

Very fine, sleek men's watch; captivating dial with mounted dart indexes and dauphin hands. Gold case (18 karats); self-winding.

**Value:** ★★★★★

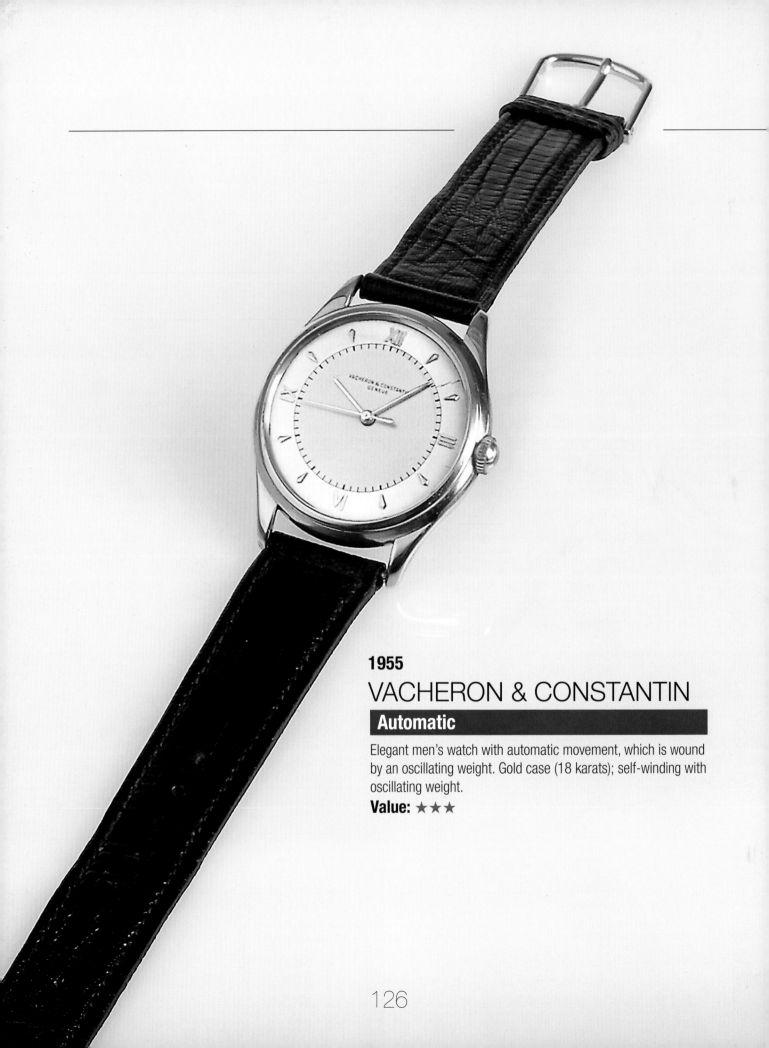

**1955**

# VACHERON & CONSTANTIN

## Automatic

Elegant men's watch with automatic movement, which is wound by an oscillating weight. Gold case (18 karats); self-winding with oscillating weight.

**Value:** ★★★

**1959**

# OMEGA

**Constellation Chronometer "Grand Luxe"**

The Constellation series established the still unbroken chronometer tradition of the Omega brand. The bracelet "à briques" is a special gem. Red gold case and bracelet (18 karats); self-winding.

**Value:** ★ ★ ★ ★ ★ ★

127

**1956**

# ROLEX

## Oyster Perpetual Tru-Beat Ref. 6556

An extremely rare variation of the chronometer wristwatch is the Tru-Beat with its eye-catching second hand. This is a mechanical movement, not a quartz watch! Stainless steel case; self-winding.

**Value:** ★★★★

## 1960
# ROLEX

**Oyster Perpetual Chronometer Officially Certified**

A special version of the typical "Oyster" with fluted bezel; comes with a deep black enameled dial. Gold case (18 karats); self-winding.

**Value:** ★★★★★

**1965**

# VACHERON & CONSTANTIN

**Automatic**

This elegant men's watch is equipped with a self-winding rotor and carries the word "Automatic" on the dial. Gold case (18 karats); self-winding.

**Value:** ★★★★

**1950**
CYMA

**Chronometer**

Elegant wristwatch chronometer with certified clock accuracy; hand-wound.

**Value:** ★

**1957**
PIAGET

**Men's Watch**

Elaborately crafted white gold men's watch with exquisite hand-wound movement.

**Value:** ★

**1967**
PIAGET

**Men's Watch**

Extra-thin men's watch in white gold; case sides adorned with Parisian horseshoe-nail pattern.

**Value:** ★ ★

**1948**
LECOULTRE

**Futurematic**

The future belongs to self-winding and the power reserve indicator; still with oscillating weight.

**Value:** ★ ★

**1958**
UNIVERSAL GENÈVE

**Polerouter Automatic**

Sporty men's watch with fine micro-rotor automatic movement.

**Value:** ★

**1960**
VACHERON CONSTANTIN

**Chronomètre Royal**

Automatic version of the model that has been manufactured (in many variations) since 1908.

**Value:** ★ ★ ★ ★

Sports Watches & Professional Gadgets

# Watches for Highly Demanding Activities

In the beginning, rugged, waterproof, and shockproof watches
were not meant for civilian use at all. But that which was suitable
for professionals was also agreeable to sporting amateurs. Beyond
their practical applications, these dependable watches also offered
their wearers the opportunity to distinguish themselves — as divers,
pilots, skydivers, or secret agents.

**1958**

# ROLEX

**Oyster Perpetual GMT-Master Ref. 6542**

The first practical wristwatch with a second time zone. It has an additional 24 hour hand, which, regardless of the new local time set by the crown, conserves the home time (in a 24 hour scheme). Early model with Bakelite bezel. Stainless steel case; self-winding.

**Value:** ★★★★★

## 1965

# BLANCPAIN

### Fifty Fathoms

Waterproof up to a depth of 50 fathoms (just under 295 feet) — this was requested by French Navy divers, who called for the development of a diving watch in the early 1950s. Blancpain won the contract. Stainless steel case with rotating bezel; self-winding.

**Value:** ★★★

**1965**

# ROLEX

**Oyster Perpetual Submariner Ref. 5513**

In the late 1950s and early '60s, the development of ultra-waterproof sports watches made a quantum leap. Rolex outfitted the professional divers of COMEX with the "Submariner," and Sean Connery made the watch famous in the film *Dr. No*. It was also coveted by non-pros. Stainless steel case with rotating bezel; self-winding.

**Value:** ★★★★

**1970**

# OMEGA

## Seamaster Professional "Ploprof" 600m

Specifically designed for pro divers, or *plongeurs professionels* in French (shortened to "Ploprof"). Based on its Seamaster diving watches, Omega developed this version, which guaranteed to be absolutely water-tight for depths of up to 1,968 feet. Stainless steel case with special crown screw and lockable bezel; self-winding.

**Value:** ★★★★

DE LA QUALITÉ EN SÉRIE
QUALITY PRODUCED IN SERIES
QUALITÄT IN SERIEN

## Portrait

# Breitling

The company **Breitling** was founded in 1884 by Léon Breitling in La Chaux-de-Fonds. From the beginning, they specialized in making excellent chronographs. Breitling wristwatches were famous for their dependability, and their endurance qualities were put to the test on the battlefield during World War I.

With this reputation, Léon's son Gaston, and later his nephew Willy, turned the company into one of the leading manufacturers of chronographs. The name always stood for high quality, robustness, and stability. In the 1930s, Breitling was considered the first choice when it came to outfitting airplane cockpits with the best chronographs. Essential for navigation, the watches had to be precise and withstand the ever increasing forces of acceleration in modern aviation.

Although Breitling didn't manufacture its own movements, assembly specialists succeeded in adapting perfectly to the needs of pilots. In 1952 the company introduced the watch that was to become a status symbol and an object of desire for pilots the world over: the Navimeter.

Its distinctive feature was a bezel with a built-in slide rule, which had two logarithmic scales that turned opposite to one another. It transformed the watch into a universal navigation instrument, long before computers took over the calculations of course, flight time, and fuel consumption. In 1962, the Navimeter experienced its first trip to space on the wrist of astronaut Scott Carpenter aboard the Mercury space capsule.

In the late '60s, Breitling, together with Heuer, Hamilton-Büren, and Dubois-Dépraz, pushed for the development of an automatic chronograph caliber.

In 1969, after a few years of research, the four companies were able to introduce the result of their teamwork: the caliber 11. For a long time the caliber 11 (along with subsequent calibers 12 and 15) was the only alternative to the "El Primero" from Zenith when it came to building automatic chronographs.

Despite this outstanding development, and although Breitling tried to adjust to the changing watch market with the production of quartz watches, the company crash-landed in 1979. After bankruptcy, the brand was taken over by Ernst Schneider, a businessman and avid pilot, and the company moved from La Chaux-de-Fonds to Grenchen. Schneider drew upon the brand's strong aviation image and firmly positioned Breitling in the luxury segment of professional chronographs.

**1973**

# OMEGA

## Seamaster Automatic 200m

Omega's diving watches are part of the Seamaster series, which are still in existence today. The design has not changed much since the 1970s. Stainless steel case with rotating bezel; self-winding.

**Value:** ★★★

**1972**

# IWC

## Aquatimer Automatic

Elegant, waterproof men's watch; rotating ring (protected by the glass cover) with 60 minute graduation and can be adjusted by the crown. Stainless steel case; self-winding.

**Value:** ★★★★

**1963**

# IWC

**Ingenieur**

The distinguishing characteristic of the Ingenieur, which was introduced in the '50s, was always its soft iron inner casing for shielding the movement from magnetic radiation. Stainless steel case with soft iron casing; self-winding.

**Value:** ★★★★

**1967**

# ROLEX

**Oyster Perpetual "Milgauss" Ref. 1019**

After completely re-casing the movement with an additional inner case, the Milgauss was immune to magnetic fields up to around 1,000 gauss. Stainless steel case; self-winding.

**Value:** ★★★★★

**1965**

# PATEK PHILIPPE

## Amagnetic Ref. 3417

This men's watch with magnetic field protection from Patek Philippe is extremely rare. Only a few of the Reference 3417 watches were equipped with a beryllium bronze escapement and a soft iron cage around the movement. Stainless steel case; hand-wound.

**Value:** ★★★★★

The hand-wound caliber 27AM-400 comes with escapement components made of anti-magnetic beryllium bronze.

## 1960

# BREITLING

### Navitimer AOPA

The superb reputation of the Breitling brand as a specialist in aviator watches can be traced back to the success of the Navitimer. Rough estimates, such as fuel consumption, cruising range, etc., could be made using its slide rule scales. Special edition for the Aircraft Owners and Pilots Association. Stainless steel case, rotating ring with logarithmic scales under the glass cover; hand-wound.

**Value:** ★★★

## 1960
# BREITLING
### Navitimer Cosmonaute

This aviator's chronograph has a 24 hour dial, which allowed cosmonauts to keep track of day and night hours during their isolation training, and even more importantly during missions in space. Stainless steel case, rotating ring with logarithmic scales under glass cover; hand-wound.

**Value:** ★★★

**1965**

# BREGUET

**Aviator chronograph "Type XX"**

Tradition-rich Breguet survived the '50s and '60s while only operating as a supplier to the French military. Stainless steel case with rotating bezel; hand-wound.

**Value:** ★★★★

**1975**

# ZENITH

## Aviator chronograph "El Primero"

Zenith manufactured a series of professional aviator chronographs for Cairelli in Rome, which were sold to the Italian military under the name "Cronometro Tipo P-2." Equipped with automatic chronograph movement "El Primero." Stainless steel case with rotating bezel; self-winding.

**Value:** ★★★

Not luxurious, but quite durable:
the basis for this Rolex chronograph movement
comes from Valjoux.

**1962**

# ROLEX

## Oyster Chronograph Ref. 6236

Rolex collectors named Reference 6036 and Reference 6236 after ski racer Jean-Claude Killy, who made this sporty chronograph famous. Stainless steel case; hand-wound.

**Value:** ★★★★★★

**1964**

# OMEGA

## Chronograph Speedmaster Professional

Before it became part of NASA's official equipment during the first moon landing, the robust Omega chronograph enjoyed enormous popularity among professionals. It was even popular among doctors, as this rare layout with pulsometer scale proves. Stainless steel case; hand-wound.

**Value:** ★★★

**1969**

# ROLEX

## Oyster Cosmograph Daytona "Paul Newman" Ref. 6262

No wristwatch chronograph has ever experienced the kind of precipitous rise that the Rolex Cosmograph has — it is worth much more today than it was then. The model pictured here with the red "Daytona" logo and square indexes is also known as the "Paul Newman" model because the actor once wore it in a Hollywood film about car racing. Stainless steel case; hand-wound.

**Value:** ★★★★★★

**1983**

# ROLEX

## Oyster Cosmograph Daytona Ref. 6265

The last Cosmograph with a hand-wound movement enjoys great popularity among collectors. Back then, though, it lived in the shadow of the automatic Daytona, which was offered at the same time. Gold case (18 karats) with bracelet; hand-wound.

**Value:** ★★★★★

**1956**
OMEGA

**Railmaster**

Hand-wound watch with extra thick dial; magnetic field protection.
**Value:** ★★★★

**1968**
OMEGA

**Speedmaster Professional Chronograph**

The official watch of NASA's manned space program; hand-wound.
**Value:** ★★

**1960**
ROLEX

**Oyster Cosmograph Daytona "Paul Newman"**

Interesting dial variation of the (at times) most coveted classic chronograph.
**Value:** ★★★★★★

**1953**
ROLEX

**Oyster Perpetual "Turn-O-Graph"**

Godfather of all "professional" wristwatches, known then as "tool watches."
**Value:** ★★★★

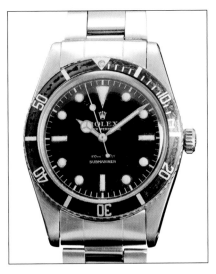

**1957**
ROLEX

**Oyster Perpetual Submariner 100 m/330 ft**

Prototype of the modern diver's watch, changed only marginally since.
**Value:** ★★★★★

**1963**
IWC

**Ingenieur**

Men's watch with an additional inner case to protect the movement from magnetic fields.
**Value:** ★★★

## Avant-garde of the 1970s
# The End of an Era

With the appearance of the first digital display quartz watches, the era of mechanical wristwatches drew to a close. Behind futuristic case shapes and colorful dials hid an outdated technology that was not able to overcome cheap Japanese quartz watches. Only a few stylistic and conceptual competitors survived the farewell to form the nucleus for a "mechanical renaissance" in the 1990s.

**1970**

# OMEGA

## Chronograph "Bullhead"

The skewed placement of a (conventional) chronograph movement resulted in a very peculiar arrangement of crowns and pushers. Stainless steel case with inner revolving bezel, adjustable by crown; hand-wound.

**Value:** ★★★★

**1970**

# OMEGA

## Seamaster Chronograph

In collector circles, this prominently shaped chronograph carries the nickname "Anakin Skywalker," after the science fiction hero from *Star Wars*. The reason for this is the mysterious luster of the bezel's tungsten carbide. Stainless steel case; hand-wound.

**Value:** ★ ★ ★

**1970**

# OMEGA

## Speedmaster Professional Mark II Racing

The larger, oval block case and lively dial lettering characterize the updated design of this classic Speedmaster Professional (the Moonwatch). Stainless steel case; hand-wound.

**Value:** ★ ★

**1973**

# OMEGA

**Speedmaster "125th Anniversary" Chronometer**

For Omega's 125th anniversary, this classic Omega chronograph was offered in a solid block case with automatic movement. Limited to 2,000 models. Stainless steel case; self-winding.

**Value:** ★★

## Portrait

# OMEGA

In 1848, the 23-year-old Louis Brandt launched a so-called "Comptoir d'établissage" (a handicraft business for assembling watch components with affiliated sales and marketing departments) in La Chaux-de-Fonds. Thanks to their attractive appearance and high quality, it wasn't long before Brandt's watches found success. The watches were also in demand abroad, and Brandt spent many weeks traveling throughout Europe.

Brandt's sons, Louis Paul and César, continued their father's active sales policy and moved the company's head office to Biel, where they set up a factory for the rapidly growing business. By 1889 it was the biggest watch factory in Switzerland.

Although the Brandts at times cooperated with other manufacturers for some of their models, the company's emphasis lay in the development of independent watches and movements. One particularly successful caliber, which through its simple construction and the interchangeability of its parts was way ahead of most other competitors' products, gave the company the name that still stands today: Omega, the last letter in the Greek alphabet. In 1903, all previously used brand names were abandoned and Paul-Emile Brandt assumed control of Omega, which he held for fifty years.

Precision was always the first priority with Paul-Emile Brandt, which is why the British Royal Air Force chose Omega as their official service watch in 1917. One year later, the US Army followed. In 1919 an Omega won the Neuenburg Observatory's precision contest for the first time. Omega's wristwatches were also successful in this popular contest, and thus the Constellation chronometer series was born.

OMEGA

Without a doubt, the most famous of all Omega models is the Speedmaster. This rather inconspicuous chronograph became official equipment of NASA astronauts due to its dependability, precision, and durability, but it had already become a legend before the first moon landing.

The brand also made a name for itself in the world of sports timekeeping. Omega was, and still is, the official timekeeper for important international sporting events like the Olympic Games. A diving watch line was also created for water sports enthusiasts, appropriately named the Seamaster. Arguably the best Omega diving watch is the model 600, also called the "Ploprof" from the French *plongeur professionel*, meaning "professional diver." The Ploprof was put to the test by the industrial deep sea diving company COMEX. Even more popular is the Seamaster Professional, which was showcased in films and on TV.

In the 1920s Omega merged with Tissot to create SSIH (Société Suisse de l'Industrie Horlogère). SSIH later merged with ASUAG (including Longines, among others) to create the SMH (Société Suisse de Microélectronique et Horlogerie) led by Nicolas G. Hayek. SMH was eventually renamed the Swatch Group.

**1975**

# HEUER

## Chronograph Monaco

A rectangular case, equipped with a newly developed self-winding chronograph movement. The Monaco is a style icon of the 1970s. Stainless steel case; self-winding.

**Value:** ★★★

**1970**

# BREITLING

## Chrono-Matic GMT

As a developmental partner of Heuer, Hamilton, Büren, and Dubois-Dépraz, Breitling also profited from designing their own automatic chronograph movement, pictured here with additional 24-hour display. Stainless steel case with rotating bezel; self-winding.

**Value:** ★★

## 1975

# HEUER

**Chronograph Calculator**

With an oversized case, space for a wide rotating bezel with slide rule function was created. Stainless steel case, rotating ring with logarithmic scales; self-winding.

**Value:** ★★

**1975**

# ZENITH

**Chronograph "El Primero"**

In 1969, the factory in Le Locle won a head-to-head race to present the first self-winding chronograph — thus the proud nickname "El Primero" for the Zenith caliber 3019PHC. It was also quite stylish. Stainless steel case; self-winding.

**Value:** ★★★

**1970**

# OMEGA

## Memomatic

The large surface case with hidden lugs earned this and similar models the unflattering nickname "fried egg." Rare version with self-winding alarm movement. Stainless steel case; self-winding.

**Value:** ★★

**1975**

# JAEGER-LECOULTRE

**Memovox Automatic "Snowdrop"**

Valuable men's watch with self-winding alarm movement in heavy gold case and "à brique" bracelet. The perfectly round shape of this watch earned it the nickname "Snowdrop." Gold case (18 karats) and bracelet; self-winding.

**Value:** ★★★★

169

**1977**

# IWC

## Ingenieur SL "Jumbo"

This classic sports watch with magnetic field protection also had to keep up with the times in the '70s. It received a bigger case with an integrated bracelet and improved protection from magnetic fields. Stainless steel case with soft iron cage. Design by Gérald Genta; self-winding.

**Value:** ★★★★

**1972**

# AUDEMARS PIGUET

## Royal Oak

When Audemars Piguet presented this stainless steel watch with integrated bracelet in 1972, watch lovers were shocked by its high price. But the concept of a luxurious sports watch made a big splash and within a short time the other big manufacturers followed. Designer Gérald Genta created a real classic, which is still produced today almost unchanged. Stainless steel; self-winding.

**Value:** ★★★★

# AUDEMARS PIGUET

Le Brassus, 1875. Jules-Louis, the youngest of the Audemars watchmaking family, started a workshop in his parents' home where he built complicated movements of the finest quality. A short time later, Jules-Louis began to have some success: the orders were flowing in and his watches were in demand in Geneva. In order to avoid any supply difficulties, he had to recruit some watchmakers from his family and circle of friends – among them was Edward-Auguste Piguet, whom he knew from school and church choir.

Since Edward-Auguste Piguet was also committed to high-quality watches, they decided in 1881 to found an incorporated company, **Audemars Piguet** S.A. From then on, they no longer made calibers for other brands. Instead, they manufactured pocket watches under their own name, usually with extra functions like repeating mechanisms.

Although the pair opened a branch office in Geneva, Le Brassus in the Vallée de Joux remained the brand's headquarters.

The old workshop, however, no longer sufficed. In 1907 they constructed a new factory, which is still the headquarters of Audemars Piguet.

In 1918 Audemars passed away, and one year later, his partner Piguet died. Their sons, Paul-Louis Audemars and Paul-Edward Piguet, took over the company. The production of the modern wristwatch presented no problem for Audemars Piguet since they had already been making small movements for women's watches.

**1957**
HAMILTON

**Ventura**

A look into the future: cutting edge design for a modern electro-mechanical movement.
**Value:** ★★★

**1965**
GIRARD-PERREGAUX

**Chronomètre HF**

Just before the outbreak of the quartz crisis, GP focused on high-frequency hand-wound movements.
**Value:** ★★

**1973**
HEUER

**Chronograph**

With orange-colored, anodized case and traditional hand-wound technology.
**Value:** ★

**1970**
CERTINA

**288**

Elaborate but conventional technology; modern case design.
**Value:** ★

**1975**
TISSOT

**Navigator Automatic Chronograph**

With polygonal-shaped case and distinctive stainless steel bracelet.
**Value:** ★★

**1975**
LONGINES

**Comet**

Hand-wound watch with time displayed by rotating discs with an arrow and a dot.
**Value:** ★

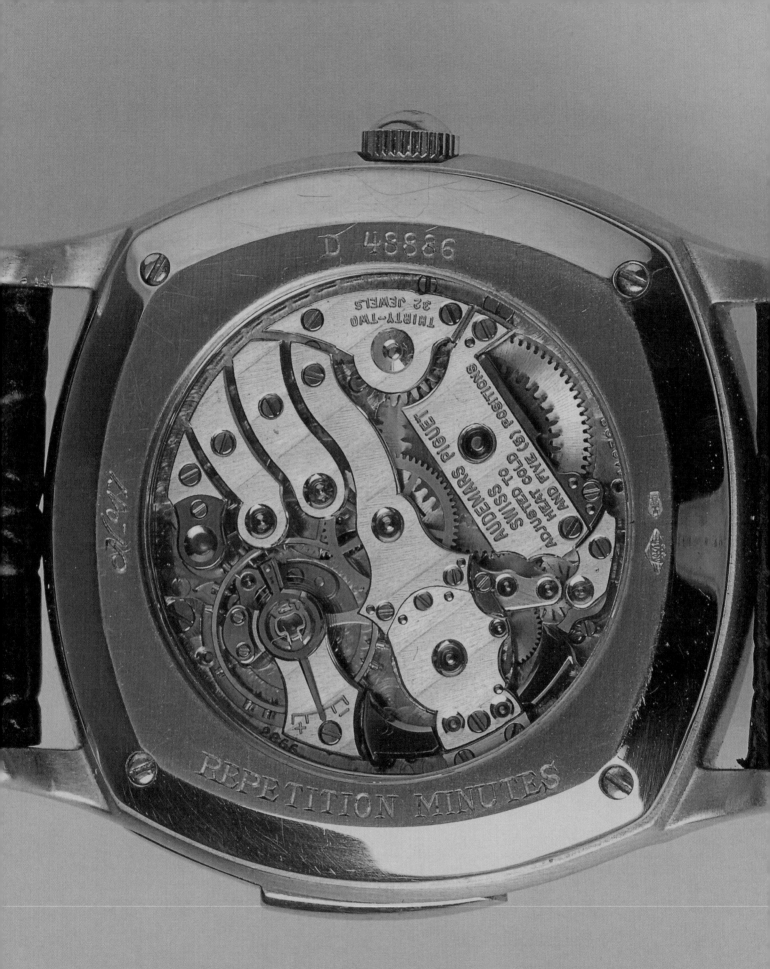

## Mechanical Renaissance

# New Masterpieces

In the '80s it looked as if traditional watchmaking had become obsolete. The Swiss watch industry was on its last legs and hundreds of suppliers had to close. With the "invention" of Swatch, a new interest in wristwatches arose. In the wake of the modern cheap watch, classic watchmakers chimed in with new products of the tried-and-true tradition and won over watch lovers who were disappointed with unambitious quartz watches. Men began to wear valuable wristwatches as jewelry, and prices exploded.

**1998**

# BLANCPAIN

## Tourbillon 8 days

Blancpain was among the first brands to stand up against the popular trend of maintenance-free quartz watches in the early '90s. With bold, complicated mechanical watches they successfully positioned themselves in the market. The tourbillon movement has an extremely long running time of 8 days. Gold case; hand-wound.

**Value:** ★★★★★

**1992**

# BREGUET

## Tourbillon

A tourbillon by the inventor of the tourbillon. In 1801 the great watchmaker Abraham-Louis Breguet patented the idea of an escapement that constantly rotates around itself. During the mechanical renaissance, his heirs revived the tradition and adorned the tourbillon wristwatch in classic 18th-century style. Gold case; hand-wound.

**Value:** ★★★★★★

**1993**

# A. LANGE & SÖHNE

## Tourbillon "Pour le Mérite"

In the early '90s, after considerable time and effort, the revived watch brand from Glashütte, A. Lange & Söhne, produced a new collection of wristwatches. This technically sophisticated tourbillon is regulated by a balance wheel with chain and fusee, and built according to old tradition. Gold case; hand-wound.

**Value:** ★★★★★★

**2005**

# VACHERON CONSTANTIN

## "Malte" Tourbillon

The tradition-rich Geneva manufacturer also returned to its glorious origins during the revival of fine mechanics in the '90s. To this day, classic complication watches and tourbillons are among the best sellers of their product line. Platinum case, tourbillon cage in the shape of a Maltese cross; hand-wound.

**Value:** ★★★★★★

**2006**

# ULYSSE NARDIN

## "Le Freak" Tourbillon

In addition to classically designed complication watches that were produced with fine craftsmanship and high expenditures, a new category of exceptionally complex concept watches emerged, which, to some extent, were made using modern materials (silicon) and modern manufacturing methods (galvanoplasty, ion etching methods). The "Freak" comes with an innovative escapement and a movement that rotates around its own axis, which serves as the minute and hour hands. Hand-wound.

**Value:** ★★★★★

**1996**

# IWC

## Grande Complication "Il Destriero Scafusia"

"The Warhorse of Schaffhausen" is the official name of this extraordinarily complex wristwatch, whose inner mechanism consists of more than 750 components. In addition to a split-second chronograph and a perpetual calendar, this watch also offers a mounted flying tourbillon and minute repetition.

**Value:** ★★★★★★

# Glossary

**Mechanical wristwatches are very complex little timepieces with numerous microscopic components. It is difficult to explain these mechanisms, their functions and intricacies, without some technical terminology. Here is a small glossary of frequently used terms in this book:**

## Automatic (self-winding)

The winding is self-actuated (by movement of the arm). This already existed in pocket watches with pendulums or sliding weights. In 1931, Rolex patented the self-winding rotor, which later became widely accepted (beginning in the 1960s).

## Balance spring

A highly elastic, precision-crafted steel spring, which, in cooperation with the pallet fork, ensures the constant oscillation of the balance wheel.

## Balance wheel

Ring-shaped component that distributes a regular beat to the movement. The balance spring and pallet fork help it maintain its balance.

## Beveling

Polishing, or finishing, the rough edges of machined movement components.

## Caliber

Another name for a movement, used to designate a specific model.

## Chronograph

A watch with a built-in stopwatch independent of the display clock that can be started, stopped, and reset.

## Chronometer

A very precise watch, whose clock accuracy is certified by an independent testing agency (COSC or an observatory, for example).

## Flyback chronograph

A special chronograph function that allows a running timer to be reset without stopping the mechanism. It also starts without any time lag on other time measurements.

## Geneva waves

The striped ornamentation on the rear surface of a movement.

## Hand-wound

The mainspring is tightened solely by the crown.

## Lugs

These fasten the strap or bracelet to the watch case.

## Magnetic field protection

Shielding a movement from magnetic radiation with an additional inner casing made of a "soft" metal (less magnetizable), usually low-alloy steel. Magnetized movement components can negatively affect a watch's operation.

## Main plate

Base support of a movement's components, which is attached to the watch case.

## Mainspring

The power source of mechanical watches, which is supplied to the movement by either manually winding the crown or by a self-winding rotor.

## Micro-rotor

Winding rotor integrated into the movement with a small diameter (smaller than the movement's diameter).

## Minute repeating

After pressing a lever, small hammers strike against thin bells and count down the hours, quarter hours and minutes with sound signals.

## Mousse

Ornamentation in the form of a pearl-like, dotted pattern, in most cases found on the dial side of a movement.

## Pallet fork

Together with the balance wheel and balance spring, forms the tiny "pulse maker" of a mechanical movement.

## Pendulum oscillating weight

See: Automatic (self-winding)

## Perpetual calendar

Mechanically "pre-programmed" calender displays, which automatically factor in the varying lengths of the months as well as the added day in leap years without having to be adjusted.

## Power reserve

Run time of the movement after fully winding the mainspring.

## Quartz

Electrical stimulation with a 32.768 Hertz oscillating quartz crystal provides the motion work's stepper motor with the proper tact. High accuracy is achieved through high oscillating frequency.

## Rattrapante chronograph

See "Split-second Chronograph"

## Self-winding rotor

See: Automatic (self-winding)

## Split-second chronograph

Comes with two second hands superimposed over one another, one of which can be temporarily stopped (to note an interval, for example), while the other continues to run. At the push of a button, the paused hand catches up with the other hand again, and they proceed on together. Resetting to zero also happens synchronously.

## Tourbillon

The escapement (balance wheel, balance spring, pallet fork) is placed inside a delicate frame, which allows it to permanently rotate around itself. This compensates for the disruptive effects of gravity.